Health and Science Essentials for Everyone

Martin A. Winkler, Ph.D.

This book is dedicated to the memory of my sister Katherine

Winkler, Martin A.
Health and Science Essentials for Everyone
Second Kindle Edition, 2020
Biotech Advisor
Lawrence KS

ISBN 978-0-9788925-5-5

Introduction

As a research biochemist and a teacher of Microbiology and Human Nutrition, I am often struck by the fragmentary and helter-skelter nature of nutritional and health information in the media and the internet. While it is important to know the latest research findings, what is often missing for many people are the basic scientific underpinnings of these areas. Fundamental knowledge of how cells, bacteria and viruses work is essential to understanding nutrition and health. Basic laws and principles of physiology, biochemistry, and genetics are introduced sequentially and concisely. Rules of thumb are needed for one's normal metabolic and physiological states. Since one's well being and happiness is influenced by the events in our physical world and comprehension of our place in the Universe, half of this book is devoted to these science topics. Chapter Two provides basic geology, meteorology, physics and Chapter Three covers modern astronomy and cosmology. The book proceeds from what we know about our body to the world around us and then to the entire Universe. I hope that this journey provides information and understanding needed to survive, thrive, and be filled with wonder at our place in the universe.

TABLE OF CONTENTS

Chapter One: Matters of Life and Health

The Rule of Three: "Three Minutes for air, three days for water and three months for food."

This rule of thumb on the time limits for going without air, water and food was given to me by my Uncle Louis Shlipak, a Doctor of Internal Medicine. The human body cannot store oxygen except in the hemoglobin in the blood and this amount is limited compared with the demands on it from cells respiring in tissues. Some mammals such as seals and whales have the protein myoglobin, which stores oxygen for much longer periods of time than three minutes. But humans and other terrestrial mammals only have three minutes or so of oxygen bound to hemoglobin in red cells in the blood. After this time, we become panicked for air and then pass out and die due to insufficient respiration in the brain. People who experience a heart attack cannot breathe and the heart ceases pumping blood. That is why is necessary to apply CPR, rhythmically pressing on the chest and blowing into the mouth (with the nose closed shut) at least ten times per minute after tilting the head back to maximally open the airway. The body is primarily comprised of water, and ample amounts are required for cells to function. The body carefully regulates the water composition of blood and organs through the kidneys, skin, and our fluid intake. When the ratio of salt and protein to water exceeds a threshold, we are dehydrated, and organs function less efficiently. The ability to cool the body decreases since sweat is formed less on the skin. After three days of no water intake, the body processes shut down, the body goes into crisis, the brain goes into a coma and the heart stops beating. Most people who are of normal weight have enough fat stored to fuel the body for a month or so during starvation. After this time, the body begins to draw down on muscle, using its protein as fuel and as amino acids. In a person of normal weight, these reserves are exhausted and then the body enters extreme starvation, resulting in death at approximately three months.

Normal pulse is in the range of 60-100 beats per minute. Pulse is an indication of physical condition in a person who is not experiencing an illness (when it may be elevated 50% or more as in a fever). A pulse less than 60 can be found in athletes, especially those who train for endurance events such as long distance running or bicycling. A sustained pulse rate in a resting person over 100 is an indication of illness or poor physical condition; a physician should be consulted.

LDL to HDL ratio should be less than 3.5 to reduce risk of cardiovascular disease.

Total cholesterol is often used as an indication of cardiovascular health. A value of greater than 200 mg/deciliter is a flag that may indicate a need for a change of diet or medication. More accurate is the measurement of the kinds of cholesterol. These are known as Low Density Lipoprotein (LDL) and High-Density Lipoprotein (HDL), along with the lipids called triglycerides. LDL, the lipid particles that deliver cholesterol to tissues and cells, predisposes to cardiovascular disease if it is highly elevated in the blood over time. The LDL particles may deliver cholesterol to immune cells in the arterial walls, causing inflammation, plaque, and ultimately cardiovascular disease. On the other hand, HDL, the particles that remove cholesterol from blood and cells, is protective. Therefore, the ratio of the LDL to HDL is a useful summary of risk. In no instance should it exceed the value of 3.5, though risk factors such as sex, age, and conditions such as Diabetes or smoking should be taken into account to access risk at lower values of this ratio. For example, a ratio of 3.0 may indicate a problem for someone with two risk factors such as smoking and obesity (risk factors can be found at www.americanheart.org).

Cholesterol LDL is mainly synthesized by the liver but consumption of saturated and trans fats elevate its levels. LDL cholesterol may lead to cardiovascular disease when its levels in blood are chronically elevated. Cholesterol is synthesized primarily in the liver and packaged into LDL there. Statin drugs block the process of cholesterol synthesis and therefore may lower the LDL levels in the blood. A lower cholesterol level increases the cells' synthesis of their surface receptors for LDL. Increased amounts of these surface receptors decrease the blood levels of LDL. Other drugs may bind cholesterol in the intestines and cause it to be excreted instead of reabsorbed. Dietary cholesterol from foods such as eggs and fatty meats can contribute to total cholesterol, as can a diet high in saturated fats or Trans fats. Trans fats, which are artificial fats made by the partial hydrogenation of vegetable or animal oils, are particularly deleterious since they both raise LDL and lower the levels of protective HDL.

Proteins are comprised of 20 building blocks called amino

acids; eight of these, called essential amino acids, cannot be synthesized by the human body and must be obtained from foods.

Most amino acids can be synthesized *de novo* by enzymes in the cells or they are made from other amino acids. However, the essential amino acids-- Methionine, Tryptophan, Isoleucine, Leucine, Threonine, Phenylalanine, Tyrosine, and Lysine-- must be obtained from foods that contain protein. Animal meat or eggs contain all of these essential amino acids, as do some plant foods such as Soybeans or peas. If the plants do not have all essential amino acids, they can be complimented with each other to obtain a complete set of amino acids. Therefore, beans and rice, or nuts and bread can form a complete meal from a protein standpoint.

Enzymes are the proteins in our bodies that perform most cellular functions. Many drugs are specific enzyme inhibitors.

For example, one of the many enzymes that synthesize cholesterol is 3-hydroxy-3-methyl-glutaryl Coenzyme A reductase (HMG-CoA Reductase). This enzyme converts HMG-CoA into Mevalonate and is an early and rate-limiting step in cholesterol biosynthesis in the liver. The Statin class of drugs binds to this enzyme and blocks it, lowering the amount of cholesterol that is made. Many other drugs have targets that are enzymes, such as anti-HIV drugs (some target viral enzymes that replicate viral DNA from viral RNA) and blood pressure lowering medications that bind to the kidney enzyme Angiotensin.

Proteins such as enzymes are complex three-dimensional polymers of amino acids that are sensitive to changes in folding and aggregation.

Heat can alter the complex protein structure, disrupting the structure to denature it, resulting in random links between proteins; this is shown when egg whites are coagulated by boiling. Proteins in the body also may bind to each other if they are altered by genetic mutation (as occurs in aging or genetic disease). Protein aggregation contributes to diseases such as Parkinson's Disease and Alzheimer's Disease. In some cases of Parkinson's disease neural proteins are made that are too long due to mistakes in their synthesis; in such cases, called "CAG repeats" (due to extra DNA containing these three nucleotides) the proteins have extra non-functional parts coded by the CAG repeats; the altered charge properties of these proteins can cause them to aggregate or otherwise interfere with neuronal function. Diseases that have been linked to CAG repeats besides Parkinson 's are Amyotrophic Lateral Sclerosis (ALS), Huntington's Disease, and Prefrontal Dementia.

All free-living organisms can be classified as Prokaryotes or Eukaryotes and Eukaryotes evolved from Prokaryotes

Prokaryotes consist of single cell organisms such as bacteria, cyanobacteria, and the bacteria called Archaea (Viruses are dependent on living cells to multiply and thrive and so are not free living). Prokaryotic cells have membranes, contain DNA without the proteins called histones and have no complex chromosome structure. They do not differentiate into specialized cells. Eukaryotes are either multicellular or single-celled and their cells have a nucleus with chromosomes containing DNA bound to histones. Eukaryotic cells also have specialized organelles called Mitochondria for production of energy, other organelles called Golgi bodies for secretion of proteins, and the capacity to differentiate into specialized cells. The protein-making cell factories of Eukaryotes, ribosomes, are larger than those of prokaryotes. All plants and higher animals are Eukaryotes, but many single-celled organisms such as yeast, protozoa such as Amoeba, and some green algae are also Eukaryotes. The marked difference between the cell walls of bacteria and the membranes of higher animals makes it possible to design drugs that interfere with their synthesis but do not usually affect animals (allergy or other sensitivity in some individuals is an exception). Antibiotics such as Penicillin or Cephalosporins are based on this strategy. Likewise, drugs such as Erythromycin or Azithromycin bind to many bacterial ribosome subunits, blocking their protein synthesis, but do not affect those of humans.

Viruses are not independent living microorganisms and require a host to reproduce them (they are obligate parasites). They are a a complex package of proteins, DNA or RNA, lipids, and other molecules; their genetic program directs the infected cell to make its protein and nucleic acid, to reproduce, and to secrete more infectious particles. They often kill the infected cell or impair its function.

To infect a host cell, a virus must bind to the surface of the cell via specific receptor proteins or glycoproteins (proteins with sugar residues attached). Once bound to the surface, it uses the cell machinery that recycles receptors to enter the host cell cytoplasm. For example, the SARS-Cov-2 (also called COV-19) coronavirus uses the Angiotensin-2 Converting enzyme receptor to enter human cells. Then the virus membrane (fatty) layer is removed, the protein coat detached, and its DNA or RNA exposed. The virus contains enzymes that replicate the

DNA or RNA. The DNA is transcribed into viral proteins, or RNA is transcribed or reverse transcribed into DNA which transcribes messenger RNA to make viral proteins. The viral DNA or RNA must be processed and packaged by viral enzymes (usually made by the host) into the viral protein coats to make new infectious virus.

Viruses can infect humans throughout the life cycle. But viruses also affect many other animal species. In fact, many human viruses (such as Influenza A and coronavirus) may have their origin in wild animals. For example, some influenza A types originate from the recombination of H and N subunits of the virus in ducks and pigs. The SARS and COV-19 coronaviruses are most similar to viruses in bats; these may have infected humans through other animal vectors. There are other viruses that infect plants, and even viruses that infect bacteria such as E. coli (these are called Bacteriophages). Infants are susceptible to Respiratory Syncytial Virus (RSV), and young children are immunized against Polio and Measles. People of all ages are susceptible to Influenza A and B Viruses, which have the capacity to recombine with each other and so may form new variant strains. For this reason, flu shots are given yearly to try to protect recipients from the current circulating strains of flu. Influenza shots, as well as Measles and Mumps vaccinations, can prevent epidemics of those viruses provided most of the population receives the shots. This is to develop what is nick-named “herd immunity", where at least 78% of the immunized population have antibodies to the virus. The discovery of the power of vaccination was first demonstrated with the control of Smallpox, which is now virtually eliminated from the world due to this method. Many vaccines contain completely inactivated (killed) microorganisms; many modern vaccines contain only the pathogen's protein that is derived from its genes in a suitable recombinant host bacterium. Breakthroughs in history were the use of vaccination to cure or prevent Rabies, then to prevent Typhoid fever. In modern times a recombinant vaccine was developed against Hepatitis B, a blood borne pathogen that infects many millions.

Chicken Pox infects children and young adults; those who are infected with Chicken Pox do not lose the virus; it ‘hides” from the immune system inside peripheral nerves. In later life, when immunity wanes, the virus may emerge and re-infect individuals, causing the painful disease Shingles. There are now vaccines against Herpes Zoster (the Chicken Pox virus) that may be taken after the age of 50 to prevent Shingles. Vaccines also are available to prevent juvenile Chicken Pox. Viruses are also responsible for Mumps, German measles, some intestinal infections, and the common cold. The common cold is caused by any of a great number of viruses, and therefore not practically amenable to vaccine prevention. Since viruses depend completely on the metabolism of their host cells, it is much more difficult to make drugs against them than against pathogenic bacteria, since a drug that blocks that cellular metabolism would also harm the host. Some drugs against HIV target the enzymes unique to HIV that enable it to process its proteins or replicate its nucleic acids without affecting the corresponding human enzymes. The few drugs available for Influenza prevent the virus from binding to their receptors on human cells via the Neuraminic acid sugar residues on those receptors.

The normal bacterial cell doubling time is about 20-30 minutes at room temperature.

Bacterial cells are smaller and simpler than Eukaryotic (higher animal) cells. Since their volume is smaller, their higher surface to volume ratio allows more rapid exchange of nutrients and waste. Since they are simpler than Eukaryotic cells they have fewer components to replicate; they multiply rapidly at warm temperatures in the presence of nutrients contained in foods or in blood or bodily tissues. The doubling time under 30 minutes means that in four hours the number has increased more than one thousand-fold. For this reason foods that contain protein or other rich sources of nutrients should be refrigerated after preparation since bacteria grow much more slowly at temperatures below 40 degrees C. After many hours in growth media, the waste products that bacteria make and their depletion of nutrients slow or stop the bacterial replication (otherwise their mass would become astronomical in size).

While many bacteria are benign or even beneficial, some bacteria such as Salmonella, E. coli H57:O157, Shigella, and H. pylori have virulence factors that cause severe disease.

The bacterial flora normally found in our intestines helps break down fiber that cannot be digested and provides several nutrients such as Vitamin K. The normal bacterial species of the human intestine, numbering in the hundreds of species, constitute what is called the Microbiome. The Microbiome, in addition to providing nutrients and digesting sugars and other foods that the intestine cannot, provides the essential function of training the immune system to what are non-pathogenic bacteria. This prevents recognition of self (called tolerance) and can reduce the risk of autoimmunity.

E. coli and other constituents of the Microbiome that escape the digestive tract into the abdomen due to a wound or injury can be highly pathogenic. Some bacteria have acquired virulence factors that enable them to resist the body's immune system and establish an infection. Examples of this are Salmonella, which takes root in the digestive tract and can cause food poisoning, or Salmonella Typhi which cause Typhoid fever.

Most stomach ulcers are caused by the bacterial pathogen *Helicobacter pylori*, as are some duodenal cancers

H. Pylori are the most widespread human pathogen, infecting at least 40 % of the human population. It is unique among bacteria that infect mammals because it produces ammonia to neutralize stomach acid, so that it may live in the harsh acidic conditions of the human stomach. It may cause the erosion of the walls of the stomach and lead to the formation of a stomach ulcer. A course of triple antibiotic therapy may eliminate *H. pylori*. Some strains of *H. pylori* have been linked to stomach cancers due to the inflammation that they may cause. To alleviate pain and further damage due to an ulcer, physicians may prescribe drugs that block the secretion of acid by stomach cells (proton pump inhibitors) or block the nerve that triggers acid secretion (H2 inhibitors such as Ranitidine or Famotidine). Another major cause of stomach ulcers is frequent use of non-steroidal anti-inflammatory drugs (NSAIDS). Drugs in this class are Aspirin and Naproxen (Aleve). These drugs inhibit cyclooxygenase enzymes that make certain prostaglandins necessary to protect and regenerate the stomach lining. Daily full doses of NSAIDs or combinations of NSAIDS (such as Aspirin and Naproxen) may inhibit these enzymes and lead to a stomach ulcer in susceptible individuals.

Antiseptics kill bacteria on the body while disinfectants and sanitizers destroy them on inanimate surfaces.

Besides antibiotics, there are substances such as acids (e.g., vinegar), alkali (soap), or surface agents (detergents), which may inhibit or block bacterial replication by modifying the cell surfaces. Antiseptics such as active iodine, alcohol and cationic detergents destroy bacteria and are compatible with the human skin (but should not be taken internally). Disinfectants such as chlorine and ozone are appropriate for killing bacteria in water. Sanitizers such as sodium hydroxide or detergents are appropriate to reduce bacterial numbers on inanimate surfaces or in manufacturing processes. Simple substances such as salt or sugar in high concentration change the osmotic environment around microorganisms so that they may burst or keep them from growing by limiting sufficient water available inside their cells for growth.

Since viruses are not alive and are simpler in composition, they are often more difficult to inactivate than other microorganisms. Many mammalian viruses such as the coronavirus are coated in a lipid membrane; dissolution or destruction of this membrane by detergents or soaps can inactivate such viruses. At least 70% ethanol is required to assuredly dissolve this membrane or inactivate the viral protein coat. Bleach solutions of at least 2% hypochlorite can also destroy the viral proteins. Of course, bleach solutions should not be taken internally, nor should 70% isopropyl ethanol ("Rubbing Alcohol").

Common Antibiotics block bacterial cell wall formation, block protein synthesis, or block DNA replication.

Bacterial cells walls are very different from Eukaryotic cell membranes in composition and formation. They contain cross-linked amino acids, unusual (D-isomer amino acids); the enzymes that form these cross links are not found in mammalian cells. Antibiotics such as Penicillin block the bacterial cross-linking enzymes, causing the bacterial cells to burst. While Penicillin will not bind to human enzymes, in a small percentage of individuals the molecule may trigger a strong allergic reaction, which in susceptible individuals can cause anaphylactic shock or even death. Other antibiotics such as Erythromycin bind to the protein-making factories in bacteria called ribosomes, which are sufficiently different from mammalian ribosomes so that the drug does not affect human protein synthesis. Another potent class of antibiotics binds to the enzymes in bacteria which replicate their DNA. These enzymes are different from ours; drugs in this class, such as Ciprofloxacin, will not harm human DNA synthesis. Bacteria develop resistance to many antibiotics. Mutation occurs naturally in bacteria and may result in bacteria acquiring the ability to destroy an antibiotic (through acquisition of a penicillinase enzyme, for example). The bacterium will multiply and pass this trait onto its progeny. Furthermore, bacteria can exchange genetic information and pass on the genes for resistance to other bacteria. Through these processes some bacteria can become resistant to many antibiotics, and this is particularly seen in some pathogenic Staphylococci or Streptococci. There are only a few antibiotics available to which bacteria are not resistant; one is Vancomycin, which is used to treat deadly infections resulting from multiple drug-resistant bacteria. For these reasons, antibiotics should only be used when necessary to fight a serious infection; use against viral infections is inappropriate, encourages antibiotic resistance in bacteria, and this resistance can be passed on by the genetic processes described above.

Warning signs of a stroke are sudden confusion, difficulty

speaking, seeing, or walking.

A stroke is a cerebral hemorrhage or alternatively a clot (Ischemic Stroke) that starves brain tissue of oxygen and stops functions in specific areas controlling movement, thought, or vision. Therefore, the sudden onset of inability to speak, to move a limb, to smile, or to see, are possible signs of a recent stroke. The signs are particularly specific if only one side of the body is affected. An alphabetic reminder of signs of a recent stroke is "STR",where the "S" stands for ask the person to smile, the "T" is for talk, making a complete sentence, and the "R" is for them to raise both arms. A sudden severe headache with no known cause is another possible sign. If these signs come on suddenly and do not go away, then one should seek medical help as soon as possible.

Aortic Aneurysms occur primarily in ages 50 to 80 and though they may take years to develop can be immediately fatal.

The Aorta is the main artery from the heart and is about one inch in diameter. An aneurysm is a bulge that causes weakness in either the abdominal section of the artery or in the thoracic section (in the chest), or there can be a dissection of the artery, which causes a separation within its walls. Men most commonly have aortic aneurysms, and they may be painless for a period of time with no symptoms. Or they may be detected by a physician during a routine examination or by imaging techniques such as X-ray, CAT scans, or ultrasonography. They may also cause severe or persistent pain in the abdomen or the thorax and this requires immediate medical attention. Thoracic aneurysms may cause coughing and wheezing or elevated blood pressure. If an aneurysm is less than two inches in diameter, it may be watched for months by imaging techniques over time, or if it is large, surgery may be necessary. The surgeon repairs the artery and may strengthen it with a mesh stent. Rupture or threatened rupture is a medical emergency. Dissections result when a tear occurs in an artery wall, allowing blood to enter the wall and create a false channel. These are three times as common in men and about seventy-five percent occur in people aged 40 to 70 years of age. Dissections almost always cause severe and excruciating pain and are repaired with surgery and a stent.

The normal calorie intake is approximately 2000 calories, but this should be adjusted depending on one's sex, weight, height and physical activity.

To maintain a constant weight, one must balance the intake of calories with their expenditure. There are resources available on the internet that can guide one in calculating one's calorie needs based on one's weight, height, age, sex, and daily exercise. The larger one is, the greater the need for calories. The older one is, generally the less calories are expended. Men generally need more calories than women of the same weight and height. Exercise counteracts weight gain by expending calories that are consumed. If more calories are taken in than expended, weight will be gained. If fewer calories are taken in than expended, weight will be lost. During energy equilibrium (equality between expenditure and intake) weight is maintained at a constant value.

Modern food processing can remove fiber, increase the concentration of carbohydrates and sugars (especially if they are added during the processing), and increase the overall speed with which the food is absorbed (as measured by the Glycemic index of the food). A high spike of glucose absorption is followed by a spike of insulin secretion that can stress the pancreas Beta cells that produce the insulin. Over time the insulin receptors can become less responsive which may lead to Type II Diabetes, where sugars are not readily absorbed by cells and can accumulate in the blood. Glucose can be a simple reducing sugar capable of chemical reaction and therefore are normally kept at tight low levels in the blood. When those levels are exceeded over a period of years, the sugars can reduce and modify (glycosylate) proteins in the eye, kidney, and arterial walls, eventually inducing disease in those organs.

The body mass index (BMI) is a way to determine obesity in individuals who are not highly muscular.

The BMI is calculated from the formula: BMI = (weight in kilograms)/ (height in meters squared). Therefore, a person who is six feet tall and weighs 200 pounds has a BMI of 29, putting him in the category of normal BMI, which is a BMI under 30. A BMI index calculator can be found on the Web at the following URL: http://www.cdc.gov/healthyweight/assessing/bmi/ Highly muscular individuals such as weightlifters or football players cannot use this formula since muscle is denser than most other tissue (except bone), skewing the formula so that a greater proportion of the weight is due to the muscle and not fat. A BMI over 30 in most other individuals is often an indication of obesity. Obesity is a risk to health for several reasons. It may increase the level of unhealthy lipids in the blood, a risk factor for heart and arterial disease. Fat cells produce more inflammatory substances (which are made from fats) and therefore a greater number of fat cells is a risk for cancer, arterial disease and stroke. The normal blood level of glucose is 70-110 milligrams per deciliter (mg/dL) in fasting individuals

A serum glucose level of over 110 mg/dl in a fasting individual may indicate Diabetes (Type I or Type II). A level below 70 mg/dl may reflect hypoglycemia. Sustained levels above 126 mg/dl are damaging to the body due to glucose modifications of nerve, organ, or arterial proteins over time. Arteries narrow due to these changes leading to decreased circulation. Diabetics are therefore more prone to foot and leg ulcers caused by nerve malfunction (Neuropathy). Immune cells such as T lymphocytes are suppressed by high blood glucose levels and therefore infections in the ulcers are common. Proteins in organs such as the kidney or in the vasculature of the retina in the eye lose function when modified by glucose on the amino acid lysine. Cataracts are more likely to occur also. A sustained high blood sugar level is also associated with abnormalities in blood proteins that carry lipids such as cholesterol. Diabetics have higher LDL and other blood lipoprotein levels and show an increased risk of myocardial infarction and stroke. Very high glucose levels indicate that insulin is not facilitating glucose entry into cells, starving them and possibly leading to breakdown of fat. This breakdown leads to ketoacidosis, a metabolic derangement that is very damaging over time (though it is the basis of some fad diets). One way that physicians detect sustained high levels of glucose in the blood is the Hemoglobin$_{A1C}$ test, which measures the percentage of modification of the red cell protein hemoglobin by glucose. While normal blood has glucose, the higher levels sustained in untreated Diabetes cause more than 6% modification of the hemoglobin with glucose residues that is an indication of uncontrolled glucose levels. There should be no significant glucose in the urine unless someone is suffering from Diabetes. Blood glucose in diabetics may be controlled depending on Type and severity. Type I diabetes requires daily Insulin injections, whereas Type II may not require Insulin, but other drugs such as Metformin (a small molecule) or Byetta (a peptide injection). Normal individuals should also try to avoid excessive blood sugar levels due to intake of an excess of simple sugars. Consistently high blood sugar levels in non-diabetic individuals have been associated with less mental acuity and greater chance of eventually developing Type II Diabetes.

Type II Diabetes is now considered "epidemic" in developed countries since the rate of its yearly increase is accelerating. A graph of its incidence with time over the years shows an exponentially increasing curve. Epidemiologists blame this alarming situation on modern sedentary lifestyles, lack of exercise and a diet laden with calorie-dense foods that are lower in fiber and overall nutrients.

There is a class of fats called Omega-3 fatty acids that are healthful since they decrease inflammation.

Omega-3 fatty acids are a kind of unsaturated fatty acid that is found in deep water fish such as salmon and sardines. It is also found in flaxseed, nuts such as walnuts, and in oils such as canola oil. Its advantage over the more abundant Omega-6 fatty acids is that the body cannot use it to make arachidonic acid, the precursor to compounds called prostaglandins that the body uses to signal inflammation. Inflammation in the arteries may lead to arteriosclerosis and heart disease. Inflammation is also involved in the formation of some cancers, such as colon cancer and stomach cancer. The enzymes that make prostaglandins are the targets of aspirin and other non-steroidal anti-inflammatory drugs (NSAIDs). These enzymes, called cyclooxygenases, are inhibited by the NSAIDs. The natural effect of Omega-3 fatty acids is to reduce prostaglandin synthesis. Regular consumption of fish such as salmon has been shown to reduce the incidence of heart disease. The recommended ratio in the diet of Omega-6 to Omega-3 fatty acids is about three to one.

Normal Individuals should consume no more than 2200 milligrams (mg) of salt per day; Individuals with high blood pressure (or a propensity to it) should consume no more than 1500 mg.

A teaspoon of table salt contains about 2000 milligrams of sodium. Our blood and many of our body components contain sodium and it is required for life processes such as sweating. We have an inborn preference for salty taste since salt was not always as plentiful or abundant as it is in the modern diet. Modern food processing and manufacturing adds excessive sodium to the diet usually because people crave the taste, though salt can also have important preservative properties in packaged foods. Modern refrigeration, freezing, and packaging usually preserves food within the expiration dates. Old-style salting or brining of meats uses very large amounts of salt to block bacterial growth and these are above that required for a salty taste.

The kidney actively conserves sodium depending on intake. Blood pressure may be elevated by consistent intake of high sodium levels in the absence of arduous work that would sweat off much of it. Individuals also vary in their response to high sodium. Some individuals have an inherited predisposition to retain salt from inherited from ancestors who lived far from salty water in the interior of continents.

Straight (grain) alcohol has empty calories and suppresses the body's ability to burn fat for fuel.

Though beer, wine, and mixed drinks have calories from carbohydrate, straight 200 proof alcohol is neither a carbohydrate nor fat. The body can burn alcohol for fuel by first removing hydrogen to make acetaldehyde and then enzymatically converting that into acetate, which is used for fuel. Straight alcohol has 7 calories per gram, which is greater than the content of carbohydrate (4 calories per gram). Fat is also broken down into acetate so the acetate from alcohol suppresses the body's ability to break down fat into acetate. The end result is that heavy drinkers may gain weight, particularly around the waist (a "beer belly") since most alcoholic drinks also have calories from sugars in them. Two or more drinks a day increase the body's estrogen level. Therefore, drinking at or above this level is a known risk for breast cancer.

People vary in their ability to metabolize alcohol. Some people have a variant of the alcohol dehydrogenase enzyme that causes it to convert acetaldehyde into acetate less efficiently. The result is facial flushing, redness, and a more pronounced inebriation. Others have a very efficient version of this enzyme, enabling them to "hold their liquor".

Carbon is the basis of all life on earth and the chemistry of carbon governs the biochemistry of life.

Carbon is capable of four chemical bonds with itself, or other elements and these bonds are often situated in a tetrahedral configuration. This enables the vast combination of compounds of carbon, hydrogen, nitrogen, oxygen, phosphorus, and sulfur that form most of the structures of proteins, DNA, RNA, carbohydrates and fats (lipids). Other elements in the same Periodic Table family, such as Silicon, can form four tetrahedral bonds, but those bonds have a greater strength, and are not as easily broken or formed, thus limiting the variety of compounds that can be made at terrestrial temperatures.

Metals are a small percentage of the weight of living cells but are essential in most living processes such as photosynthesis, respiration, enzyme reactions and immunity.

Metals account for less than a few percent of the body, but their roles are diverse and at the heart of many of the catalytic reactions of biochemistry. Iron is found in the hemoglobin molecule and enables it to bind to oxygen or release it into tissues. It is also part of the mitochondrial enzymes that break down carbohydrates with oxygen to provide energy. Manganese is the key part of the chlorophyll molecule that drives all photosynthesis in plants or algae. Zinc is an essential component of hundreds of enzymes in the brain, immune cells and testes. Copper is also an essential cofactor for many enzymes. Cobalt makes up the active center of Vitamin B12.

Heavy Metals are the group name for Mercury, Lead, Cadmium, Nickel, and other toxic elements that may accumulate in the body and cause problems by binding to proteins and enzymes.

These elements may bind very avidly to sulfur containing amino acids as well as some other amino acids and cause inactivation of essential proteins such as enzymes in the nervous system and other organs. Mental retardation and other nervous system disorders are possible in children if their mothers ingested significant amounts of mercury or lead (in the micrograms range per day). Nickel may cause contact dermatitis, allergic reactions and pulmonary disorders as well as cancer. Cadmium may cause a variety of GI disturbances, lung disorders and mental disorders.

Mercury is found in coal, and the burning of coal is the most significant cause of mercury pollution in the atmosphere and in the ocean. Mercury accumulates in the tissue of small fish and is concentrated further in their predators, large ocean fish such as Albacore Tuna, Sharks, and Mahi-Mahi. Pregnant women should not eat these larger fish species and should also limit ingestion of other ocean fish.

Modern electronic devices such as computers are hazardous sources of toxic heavy metals. Nickel-Cadmium batteries should never enter the waste stream nor should electronic components from the devices since they may contain cadmium, nickel, and other hazardous elements. Mercury is found in compact and regular fluorescent light bulbs, and old- style electronic relays. These should be recycled at authorized facilities.

The basic unit of heredity is the gene, which usually codes for an essential protein, and there are two gene alternatives (alleles) on each chromosome pair.

A gene is the segment of DNA on a chromosome that codes for a protein or sometimes for a structural RNA in the cell. The information of the gene is transcribed by enzymes called polymerases into messenger RNA (mRNA). The mRNA moves into the cytoplasm where it provides a template on Ribosomes for the synthesis of proteins. That process uses Transfer RNAs to bring amino acids to the mRNA coding the nascent protein on the ribosome, which catalyzes the incorporation of the amino acid into the growing protein chain.

Genetic information is transferred to successive generations by genes in the sexual gametes. During the process that forms sexual gametes, one of the two alleles is chosen to be included in the gamete at random. A gene is either dominant, in which case it is fully expressed and functional in the cell, or it is recessive, in which case it is masked by a dominant gene. Recessive genes, which do not show if a dominant gene is present, can assort to give two recessives which then show in the phenotype of the double recessive. This explains why parents both with brown eyes can have a child with blue eyes (the blue eye is coded by a recessive gene). Blood types are another example of dominant or recessive genes. Types A and B blood markers are dominant, and Type O is recessive. Rh negative blood type is recessive and Rh positive is dominant. Therefore, a person with an O type marrying an A or B type will have either A or B type children. With Rh factor, a mother who is Rh negative married to an Rh-positive father may have a child who is Rh positive (especially if the father does not carry a recessive gene for Rh negative). This may cause the mother's antibodies to react against the child. To overcome complications from the mother's antibodies, a blood transfusion for the child or other treatment may be necessary at birth. There are now antibody therapeutics that can be given to Rh negative mothers expecting an Rh-positive baby; the antibodies eliminate the foreign Rh positive glycoproteins from the mother's blood, preventing harm to the baby.

There are about 23,000 genes on the 23 human chromosomes.

This number of genes is much lower than what was initially suspected by scientists. However, many genes are modified by enzymes (methylation for example) and alternate splicing of messenger RNAs from many genes in different cells or during development (as in the immune system). Alternate splicing makes it possible for many different proteins to be synthesized from the same gene. Gene expression is also controlled by intervening regulatory elements with multiple effects on gene activation or suppression. Through splicing or regulation, cells get a big "bang for their buck" out of this surprisingly low number of total genes. The genes are lined up on the chromosomes and their alignment is normally fixed in an adult cell chromosome.

Gene alleles between two sister chromosomes "cross over" frequently on the chromosomes during meiosis (sexual gamete reproduction) so that only different genes that are closely situated together will show strong genetic associations. When two different genes are closely situated ("linked") they will be found together in different generations. An example of this is the histocompatibility gene B and predilection to Ankylosing Spondylitis (a kind of arthritis). However, most genes lie on different chromosomes. For example, eye color and ABO blood type are each coded by different genes on different chromosomes and these assort randomly so that one cannot associate one's eye color with blood type. Also, for these reasons, ABO blood types are not associated with skin color or height. This process of gene sorting ensures great diversity of genotypes in the human population, providing many possibly advantageous individuals who might survive times of famine, microbial epidemics or other crisis

Chromosomes contain considerable DNA that does not code for active genes.

The human genome contains more than three billion base pairs of nucleotides of DNA. However, only less than 2% of that directly codes for proteins. Much of the rest of the non-coding sequence is structural, of unknown function, or codes for ancient genes called Pseudogenes that are damaged so that they cannot usually be expressed. Pseudogenes are molecular fossils of our evolutionary past and are not under the evolutionary pressure of functioning genes. Therefore, they accumulate random mutations at a greater rate then functioning genes.

Normal eukaryotic genes contain stretches called exons that code for protein separated by sequences called introns that do not code for protein. During messenger RNA transcription the intron sequence is spliced out. The final mRNA coding for protein makes its way to the Ribosome where it is translated into protein. Many pseudogenes do not contain start or stop codons to indicate where transcription should begin. Others have stop codons in the middle of what would be an exon, or they may lack promoter sequences to begin the transcription process.
Humans have hundreds of pseudogenes for the Olfactory Protein Receptors (OPRs), proteins whose normal counterparts enable us to detect odors. Mice have fewer of these pseudogenes and more of the normal OPRs, reflecting their greater reliance on the sense of smell and our greater reliance on color vision.

Cancer is a diverse set of diseases that have in common uncontrolled cell growth.

There are more than a hundred different kinds of cancer, and they are diverse. For example, testicular cancer is very different from skin cancer. Tumors can be either solid (e.g., a lung tumor) or disseminated in the blood (as in Leukemia or Lymphoma).

The most common cancers are lung, colon, breast, prostate, and skin. These all may be treated successfully if the tumor is detected early before it can spread to other tissues.

Mammograms may detect early breast cancer. X-rays may detect lung cancer. Blood tests and physical examination may detect prostate cancer. Skin examination is useful in screening for skin cancers. For melanoma, the deadliest of the skin cancers, the ABC rule should be followed, which specifies that melanomas may have a larger **A**rea, irregular **B**orders and a number of **C**olors in them.

Most cancer is due to mutation of cell genes that control or activate cell growth and only about fifteen percent of all cancer is due to an inherited predisposition.

Most cancer is caused by somatic mutation, which is an alteration in the genes in body cells such as lung or colon cells (as opposed to mutation in the germ cells which give rise to eggs or sperm). These mutated genes are then passed on to the progeny of the mutated somatic cells. The most common causes of somatic mutation are environmental agents such as smoke from tobacco, radiation (either natural from cosmic rays, Radon in the ground or man-made radiation from X-rays), or from substances created in the cooking or smoking of foods. Cigarette smoke, heavily charred foods or home-smoked foods contain substances called carcinogens that are created from the action of high heat on the proteins and fats in the food. Carcinogens modify DNA, damage it, and if the damage is in a gene controlling cell growth, then they may cause changes that lead to cancer. Many carcinogens act on the tissues that contact them first, such as esophagus, lung, stomach, or colon. Other carcinogens require the liver to convert them into their active forms before they are active in the blood stream. Vegetables such as broccoli or others in the cruciferous family contain substances that cause liver enzymes to be less efficient in the liver conversion to carcinogens. Intake of vegetables such as these may lower cancer incidence. Another cause of somatic mutation is bodily inflammation from immune cells or fat cells that produce over time an excess amount of active chemical substances that normally are used by the body to kill bacteria and other microorganisms. Some chronic infections such as those due to H. pylori are associated with cancer such as stomach cancer. Substances such as Asbestos can directly activate inflammation in cells that may lead to the rare cancer Mesothelioma.

Mutations in cancer often are either in genes for proteins that suppress cell growth or in genes for proteins that are involved in growth activation.

The body has an intricate system of proteins that control the initiation and progression of cell division. These proteins and the genes that code for them constitute a system for growth control and therefore suppress tumor formation. These are the largest class of genes knows to be mutated in cancer. The gene for Retinoblastoma, an inherited childhood cancer, is in this class and was the first tumor suppressor gene characterized. Other cancer associated genes code for cell surface proteins that bind growth factors and trigger cell growth. An example of this is the Her2/neu gene that codes for an epidermal growth factor receptor. If Her2/neu is mutated to be in the "on" state and signals continuously, then a breast cancer growth may result. The two classes of genes, tumor suppressors and growth factor receptors, provide a means of either "putting the brakes" on cell division, or switching it on continuously.
The immune system constantly scans for cells that have

been transformed to have uncontrolled growth; these cells can be eliminated by T cells or Natural Killer cells of the immune system. Tumor cells often evolve to express surface markers that mask them to the immune system by indicating to immune cells that they have the properties of normal cells. Pharmaceutical companies have developed antibody therapeutics that cover these tumor markers and therefore signal to immune cells that the tumor cell should be killed. These so-called immunotherapeutic drugs have shown efficacy in treating Melanoma and lung cancer, for example

Folate and Vitamin C are two water soluble vitamins that are protective against some cancers and other diseases.

Vitamin C is added to many nitrite-preserved meats to prevent the formation of Nitrosamines, cancer causing compounds formed when amino acids from protein react with nitrites in the stomach. This vitamin will also benefit in the diet when vegetables containing nitrites are consumed with protein foods. Folate is required for the synthesis of blood cells, for proper brain functioning, and during development of fetal cells. Pasta and bread are now required by law to be fortified with folate to decrease the incidence of the birth defect Spina Bifida, which is linked to low folate consumption during the first trimester. Folate is used in the synthesis of DNA; low folate intake has been associated with increased incidence of some cancers.

Vitamin C does not prevent or significantly alleviate the common cold.

It is a common misconception that Vitamin C in large doses alleviates or prevents the common cold. While Vitamin C is essential to support the general health of the immune system, it does not afford protection against the many viruses that cause the common cold. A number of large clinical trials showed that Vitamin C doses were no better than the placebo in relieving or preventing common colds (Anderson , T.W. et al, Canadian Medical Association Journal 107:503-508, 1972). The reason that this myth is so widespread is due to the efforts of eminent Nobel-prize winning chemist Linus Pauling. On theoretical grounds he believed it to be a cold preventative, but never did the appropriate clinical trials needed to test the validity of his idea. Dr. Pauling was a gifted theoretical chemist (and worked successfully to ban atmospheric nuclear bomb tests) but he did not perform statistically valid, controlled population studies that were needed to rigorously evaluate this idea. Once they were done, it was clear that there was no benefit of Vitamin C compared with placebo, but the trials had taken years to complete. By then this erroneous idea had spread greatly, aided by the commercial success of the suppliers of mega doses of this inexpensive vitamin, which Dr. Pauling had advocated.

There are rigorous statistical requirements for minimum population size in any clinical trial or health study.
Our bodies are extremely complex, and it is difficult to arrive at accurate information about medications, nutritional supplements, or medical treatments from anecdotal evidence (which is studying just a few people or making a few observations). This is commonly reported in the newspaper when they say, "studies show". and the studies cited are less than 30 people, for example. For medications, nutritional supplements or health treatments, the numbers of people needed for an accurate study are most often in the hundreds or thousands. The facts about nutrition were found through careful experiments or clinical trials, designed, executed and analyzed according to rigorous statistical requirements. These are then published in credentialed primary reference sources. Credentialed references mean that the author(s) have an advanced degree in the subject and are members of accredited national or international professional societies in that field.

Anyone regardless of qualifications can buy a web domain and put up a web site. Therefore, health or nutritional information found on the internet may need to be further validated if it is commercially motivated or the authors have ulterior or selfish motives to promote their claims.

Plentiful antioxidants are an essential part of a healthy diet.

During cell metabolism of oxygen highly reactive intermediates called free radicals are generated. These are singlet oxygen, hydrogen peroxide, oxidized lipids and other reactive compounds. These reactive oxygen species are also generated during inflammation, especially by mammalian immune cells (Neutrophils) when they kill foreign microorganisms. The body's cells contain enzymes such as Superoxide Dismutase, Peroxidase and Catalase to deal with free radicals, but the efficiency of these enzymes is not one hundred percent activity, they are not always at the same site as free radicals, and their activity decreases as we age. Free radicals can damage DNA causing mutations in cells and they can alter proteins, decreasing or blocking their activity. Oxidized fats which are found in fried foods can also damage cells and blood proteins. There are substances in many foods and beverages which scavenge or neutralize free radicals and oxidized fats. These are called antioxidants. Tea, coffee, and chocolate are rich in antioxidants, as are some foods such as mushrooms and broccoli. Vitamin C is an antioxidant that is found in many fresh fruits and vegetables. Chocolate, especially dark chocolate, is a potent antioxidant that can lower blood pressure and increase blood flow to the brain significantly. In common with tea and coffee, chocolate also contains some caffeine. Caffeine improves mental performance and alertness but consumed in excess or too close to bedtime can interfere with sleep.

The essential amino acid Tryptophan as well as carbohydrate and Vitamin B6 are required for synthesis of the brain neurotransmitter and mood regulator Serotonin.

A key regulator of mood in the brain is a neurotransmitter called Serotonin. It is essential for keeping an even mood and also for sleep. Serotonin synthesis is catalyzed by brain enzymes using the essential amino acid Tryptophan as starting material. But the blood brain barrier excludes many essential nutrients such as Tryptophan and there are special transporters in the brain to bring in Tryptophan. It turns out that these transporters also require carbohydrate (glucose) to function, as well as Vitamin B6. Some fad diets such as the Adkins Diet advocate strictly limiting carbohydrate consumption (though carbohydrate is an essential nutrient). One reason that this diet cannot be maintained over a long period of time is its possible effect on mood through limitation of carbohydrates. One of the symptoms of Pyridoxine (Vitamin B6) deficiency is nervousness and sadness which can be linked to its role in Serotonin synthesis. There is more Serotonin in the brain after a high carbohydrate meal. It is also why a meal high in Tryptophan (such as Turkey) with carbohydrate (dressing) will make one sleepy.

Our Digestive tracts evolved to require fiber for proper functioning.

Animals and humans through millions of years of evolution consumed substantial amounts of fiber daily. Only in recent modern times have humans consumed processed foods that are relatively low in fiber. Fiber stimulates the peristalsis of the large intestinal tract, keeps its muscular tone, and helps to prevent hemorrhoids, diverticulitis and colorectal carcinoma. While infants and some elderly people have reduced fiber needs and have less tolerance for larger amounts of fiber, most children and adults need around 30 grams of fiber per day. Fiber is comprised of indigestible complex carbohydrates in foods such as whole grains, fruits, and nuts. Fiber slows the absorption of sugars and other carbohydrates, lessening "spikes" of glucose in high carbohydrate meals. This also decreases spikes of insulin and therefore is particularly valuable to diabetics or those with episodic hypoglycemia. Fiber in meals along with adequate hydration assists in weight control since in provides a full feeling without adding calories.

Our Immune system consists of T cells, B cells, accessory cells and antibodies and other proteins that recognize and destroy microorganisms, viruses, and infected body cells.

All mammals have a very sophisticated system of different kinds of cells and proteins that develops from birth to recognize non-self living organisms in the body and, if necessary, to destroy them. Macrophages are cells distributed throughout the body which recognize and engulf foreign cells. They may kill some microorganisms or activate other immune cells. Dendritic cells are found in the lymph nodes as well as the liver and other organs and they also bind and trap foreign cells. T cells receive signals from these cells and coordinate the marshalling of the immune system against an infection or transplant tissue. T cells have the capability to destroy a foreign cell, as do macrophages and NK ("Natural Killer") cells. B cells recognize bacterial, viral and other non-self molecules and produce complex proteins called antibodies. Antibodies have the capacity to bind exogenous proteins, neutralize viruses, and further activate the immune system. Other proteins called Complement are activated by some bound antibodies; complement can lyse a bacterial cell. Highly mobile blood cells can be activated by the immune system to cause them to pass into tissues where they secrete active oxygen and chlorinated molecules which are toxic to microorganisms.

There is a natural variation between individuals in their specificity against foreign molecules, and also a variation in the capacity to mount an immune reaction to microorganisms. Some people do not recognize the Tuberculosis bacterium, Mycobacteria Tuberculosis, as well as others, and therefore are more susceptible to it. There is also variation in the ability to fight the bacterium that causes Hansen's disease (Leprosy).

Autoimmune disease can result when the Immune system accidentally directs its killing capacity against self.

The immune system, like a powerful army, can sometimes accidentally cause it to attack the host body. This is like "friendly fire" in the military. The immune system has built into it regulatory cells and molecules that cause it to "learn" not to attack self. This process is called tolerance and involves the "education" of T cells and B cells during early development of a mammal before adulthood. This process also occurs in birds and is one of the reasons they have a special organ called the Bursa of Fabricius, which takes in foreign material ("dirt") and exposes it to the developing immune system to stimulate it. Animals raised in axenic environments (bacteria-free environments) are deprived of this stimulation of the immune system and may succumb to infections later. Some bacteria have molecules that look like our own; when the immune system makes antibodies against them, they may bind to self and cause disease. Rheumatic fever is an example of such an autoimmune disease. It occurs when a certain strain of Streptococcus infects some individuals. These individuals make an antibody against the Strep bacterium and the antibody also binds to the heart muscle. Severe heart disease can result from Rheumatic fever; this is one reason Strep infections are treated aggressively with antibodies after rapid diagnostic tests indicate throat infection. Other autoimmune diseases are caused by the immune system reacting to tissue in a joint (Rheumatoid Arthritis), nerve sheaths in the brain (Multiple Sclerosis), RNA in skin and organs (Lupus Erythematosus), and the tissues of the colon (Ulcerative Colitis).

All Drugs are either chemical compounds ("small molecules") or macromolecules such as proteins.

The drugs that inhibit enzymes for cholesterol synthesis, such as Mevacor (for high cholesterol), are chemical compounds similar to the substrates of those enzymes, but function by blocking or not allowing the enzyme to perform its functions. These drugs are usually in the form of pills or simple solutions. Drugs that replace lost function, such as Insulin for Diabetes, are proteins that must be injected subcutaneously or intravenously. Other large molecule drugs are proteins called recombinant antibodies that are used to treat diseases such as Rheumatoid Arthritis, Crohn's Disease or Psoriasis. Recombinant Monoclonal Antibodies are also available to treat blood cancers such as Lymphoma or Leukemia, or solid tumors such as breast cancer or colon cancer, or chronic disease such as Macular Degeneration.

The rate of cooling of an animal is proportional to its surface area but the heat it generates is proportional to its volume.

The ratio of an animal's surface area to the total volume (or mass) is a measure of how quickly it can cool itself; the higher this ratio the more efficient cooling will be. The larger an animal, the smaller this ratio; in general, large animals can stay warmer and need to generate less heat to stay warm. When applied to humans, the formula simply says that people who have a higher body mass index will often require more cooling in summer and be warmer in winter. The relation between energy requirement and size is why there are no mammals smaller than a mouse. A mouse has a heart rate over 200 beats per minute and must eat one-half its weight in food each day to support its metabolism and keep a constant body temperature; an even smaller mammal than a mouse would have huge energy requirements. Some reptiles, amphibians, insects and other small non-mammals have a metabolic rate that is more susceptible to the influences of the surrounding environment since their temperature is not regulated by an internal thermoregulation system.

The normal short term digit memory span is 7 characters.

Short term memory can be measured by digit span or recall of a set of random numbers given at one second intervals. For example, the numbers 9-4-7-3-8-3-5 remembered correctly, but not a longer set, is a digit span of seven. From ages seven to adult, the average digit span is seven plus or minus two. For younger children, the span roughly corresponds to age. For this reason, we usually must write down complete phone numbers with area codes, which have a digit span of 10. For letters, the span or chunk remembered in a short time is shorter, around six, and for words, it is around five.

The average human with no special aid or training can tolerate sustained forces of 5 g force downward, 2 g force upward, or less than 12 g force horizontally.

The force of acceleration is measured in g's or multiples of

Meters per second per second. Its effect on persons depends on the direction of the force. Pilots and other trained individuals wearing "g suits" can tolerate higher values than these. A force directed so that the head is pointed in the direction of the force (such as hanging inverted or spinning in a circle with the head at the tip of the radius) is least tolerable since it draws blood into the brain and eyes, where delicate capillaries may be ruptured. The force downward is more tolerable since it directs the blood to the feet; the force horizontally (perpendicular to the spine) is most tolerable since the blood remains in the central core of the body. The horizontal axis force also varies somewhat whether it is forward ("eyeballs in") or backward ("eyeballs out") since the latter fills the eyeball capillaries with blood and may rupture them. The backward horizontal limit is 12 g and the forward horizontal limit 17 g.

The Earth did not always have an oxygen atmosphere and the oxygen we breathe is a result of bacterial fixation of carbon dioxide billions of years ago, as well as photosynthesis by plants.

Several billion years ago there was very little oxygen in the earth's atmosphere. Some bacteria and algae then evolved photosynthesis, the biological process that converts sunlight into energy to drive the fixation of carbon dioxide from the atmosphere and water into cell material with the release of oxygen gas. This activity went on for hundreds of millions of years culminating in the evolution of higher animals and also microorganisms that could utilize the oxygen for respiration. Higher animal (Eukaryotic) cells captured (or cooperated with) microorganisms capable of respiration and these are now what are called the Mitochondria—organelles in cells that specifically use oxygen to burn carbohydrate to provide energy for the cell. Interestingly, mitochondria have their own DNA, and it is like that of prokaryotes, not eukaryotes.

Diamond and Graphite, respectively, are the Hardest

and softest minerals.

What is most interesting is that both diamond and Graphene are both comprised completely of carbon. Carbon in diamonds is bonded through bonds forming a tetrahedral configuration in a crystalline matrix that makes it exceedingly strong; carbon in graphite is bonded together in planes that form sheets. The graphite sheets can slip past each other, causing the exceptional softness of graphite as found in a pencil lead. An individual plane of carbon in graphite can be isolated and is called Graphene. This is the thinnest material known and is remarkable in several other respects. Electrons in Graphene have less mass than they do in copper or other metals, and so move much faster. Graphene also exceeds even steel in its Young's Modulus, a measure of strength and toughness per unit weight

Heat capacity is a measure of the ability of a substance to hold heat.

Heat capacity is the amount of heat applied to produce a unit change in temperature. It is measured as specific heat, or units of calories per gram of a substance. Water holds one calorie per gram, and for land it is about a third of this depending on its moisture and composition; air has even less specific heat. For this reason, areas of land near the ocean have more stable temperatures than areas in the center of the continent, hence the phrase "continental climate". The thick adobe walls of a southwestern house have high water content due to their clay and therefore take longer to heat up during the hot desert summer. In winter, they hold heat longer than if they had no moist clay.

For most chemical reactions that proceed with reasonable speed, the reaction rate increases two to three times for every increase in temperature of ten degrees Centigrade.

This rule of thumb can be used by photographers developing and fixing photographic film and prints. The photography wet methods are chemical reactions in water of reagents such as sodium hypochlorite and sodium sulfate at significant concentrations. With such reactions, there is a reasonably rapid rate to completion, i.e., the reaction is over within a couple of hours at most. This rule is derived from the Arrhenius equation, which specifies that the rate constant of a chemical reaction is proportional to the Arrhenius constant times the mathematical constant *e* to the negative power of the inverse of the Temperature in degrees Kelvin and the gas constant *R*. So a plot of log k (the reaction constant) versus the absolute temperature (degrees

Kelvin) is a line with a negative slope. For reactions that proceed reasonably fast, the Arrhenius constant is high enough for this rule of thumb to hold. This rule explains why many chemical reactions are facilitated with heat. However, since exothermic reactions also release heat, a chemical reaction that is heated often must be conducted carefully with cooling in a fume hood or other protected enclosure.

The temperature lapse rate with altitude is around 2/3-degree C. per every 100 Meters (3.6 degrees F per 1000 feet altitude).

The lapse rate is the change in temperature with altitude and is an important measure when predicting weather changes. Lapse rate can be measured with radio reporting instruments under a balloon; its value varies with terrain, geography and weather conditions. If the lapse rate is 1 degree or more per 100 meters, then the air is unstable and a parcel of air (or a balloon, for example) will tend to keep rising if in motion, or even sink if propelled in that direction. Unstable air is found in hot deserts or when low- pressure fronts move through a region. If the lapse rate is less than one degree C. per 100 meters, then the air is stable and will resist rising or falling. Such conditions occur near places that get ocean breezes or are surrounded by a large body of water.

Outdoor temperature should be qualified to consider wind speed or the ambient humidity.

Moving air removes heat from one's body and chills it if cold. Therefore, meteorologists have developed the wind-chill scale which combines temperature and wind speed. For example, a person at 50 degrees F. in a wind of 25 mph a wind chill below freezing. Humidity makes warm air feel warmer since it prevents evaporation from the skin. At cold temperatures the moisture in the air removes heat more efficiently than dry air, further chilling the body. Therefore, humid cold air feels colder than dry cold air.

"Mares tails" and high Cirrus clouds are signs of weather change.

If one sees high, long wispy white clouds in a relatively clear blue sky, they are an indication of a weather change within 48 hours. Some of these Cirrus clouds are called "Mares

Tails" from their appearance. They indicate that high velocity winds carrying colder air are present at high altitude. This can be interpreted as the leading edge of a cold front or other weather front making its way into the area in the next day or two.

If one sees a ring around the moon this is also indication of moisture at high altitude carried by colder winds. Such a ring is an indication of a change in weather and possibly a cold front in winter.

El Nino is a warming of the waters every 4-7 years in the eastern Pacific that can impact temperature and precipitation in many areas of the United States.

El Nino is the nickname given to the El Nino- Southern Oscillation (ENSO). It received the name "El Nino" ('The Child") since it causes dramatic decreases in the fish catch around Christmas time in the Pacific Ocean near Peru and other parts of South America. Normally the water 50 meters below the ocean surface off the coast of Peru is cold due to upwelling of cold currents. But in El Nino, warm water piles up in the eastern Pacific and the upwelling of cold currents is suppressed, decreasing the fish feeding there and the possible catch. The warming of waters in these ocean regions causes an increase in rainfall or snowfall in the southwestern U.S., leading to possible flooding and avalanches, as well as warmer winters overall in the Midwest. In Australia and Indonesia there may be droughts and high winds. La Nina is a reversal of this oscillation, and it is caused by colder waters in the western and central Pacific. La Nina may cause increased storms and rainfall in the Northwestern U.S. and colder temperatures in the Midwest

The Jet streams are very fast streams of air at high altitude that influence weather and aviation.

There are two jet streams in the Northern Hemisphere and two in the Southern Hemisphere. The stronger one of each pair is called the circumpolar jet and the weaker one is called the subtropical jet stream. They consist of streams of air moving hundreds of miles per hour at altitudes from 20,000 to over 30,000 feet. The most predominant effects are seen in the Northern Hemisphere, as the ones in the Southern hemisphere circle around the Antarctic Ocean and have less of an effect over land. They were discovered by a Japanese Meteorologist observing balloon flight around Mount Fuji; this discovery was not released to the world at that time and the Jet Streams were re-discovered by fliers before and after World War II when their dramatic effects on wind speed were observed. Increases of more than 100 miles per hour were seen in transatlantic flights to Europe. Conversely, slowing of airplanes' apparent air speed was observed in flights to the west. The jet streams are heavily influenced by either El Nino or La Nina weather events. El Nino pushes the Northern Circumpolar jet to the south, bringing increased storms to the Southwest U.S. and fewer storms and more warming to the Midwest in the winter. La Nina causes the Jet to move North and brings strong storms to the Northwestern U.S. during winter and hot, dry summers to the Midwest.

Jet streams are not limited to the earth but occur on other planets where heating and cooling of the atmospheric gases interacts with the rotation of the planet. For example, Jupiter has numerous jets as evidenced by its striking bands.

The rate of global warming accelerates at 450 ppm atmospheric carbon dioxide.

Carbon dioxide in the atmosphere is the primary cause of what is termed the "Greenhouse Effect". This gas absorbs solar energy in the atmosphere but inhibits its re- radiation into space. Methane also has this property but is found at much lower concentrations in the air (1.8 ppm). Man's energy burning activities on earth—principally coal burning, gasoline and other hydrocarbon combustion, and forest fires are the primary reason that the carbon dioxide concentration in the earth's atmosphere has reached 412 ppm in the year 2022. The sustained higher concentration of carbon dioxide has produced dramatic evidence of global warming. Over 250 extra billion tons of Greenland ice melted in the years 2005-2008. Historically unprecedented melting of arctic ice seas also recently occurred. If this trend continues, by the end of the twenty first century the ice may disappear from the North Pole and polar bears will have no ice flows to live on--their survival will be in danger.

If the trend of global warming continues the end of the century could witness significant elevation of sea levels, inundating many major cities and urban areas. At 450 ppm carbon dioxide the warming accelerates. Thawing of arctic ice releases methane as well and it is also a greenhouse gas. Substitution of non-carbon consuming sources for energy is needed in addition to decreases in deforestation. Forests and other plants remove carbon dioxide from the air and substitute oxygen through the process of photosynthesis.

The seconds after a lightning flash should be counted and divided by five to determine the approximate distance of the lightning bolt in miles.

The speed of sound in air at 22 degrees C. is 772 miles per hour, or 0.21 miles per second. So, in five seconds sound has traveled 1.07 miles. Since light travels at 187,000 miles per second the sight of the lightning flash is virtually instantaneous. By dividing the time in seconds by five we can determine the approximate distance in miles.

The speed of sound increases somewhat with the temperature and with the composition of the medium.

The speed of sound in water at 22 degrees C. is 3367 mph, over four times faster than the speed in air. In salt water the speed is even 6% faster. The velocity of sound in steel is fifteen times that in air and in diamond it is 36 times as fast! The velocity of sound in helium, which weighs an eighth of the average weight of air, is 2.8 times faster than that that of air. For a gas, the formula states that the velocity is inversely proportional to the square root of the mass of the molecule conveying it. Since the size of the voice box is fixed, a person inhaling helium then speaking has a very high-pitched sound, due to the faster velocity of the waves in helium. The increased frequency is because the wavelength is fixed by the voice box and frequency equals velocity divided by wavelength.

Temperature increases about 2-3 deg. C. with each 100 Meters depth into the Earth.

This gradient of increasing temperature is found around the earth, with a more rapid rise to 1000 deg. C. occurring in tectonically active regions beneath the earth that are responsible for volcanism. In areas that are not tectonically active, the rise in temperature is more gradual. Due to the temperature rise, the mantle of the earth is molten, and this liquid rock, called magma, can rise through volcanically active areas of the crust. The temperature does not continue to increase all the way to the center of the earth, but reaches a maximum temperature around 25,000 deg. C. The temperature rise in the earth is due to the decay of radioactive elements. It is thought that these radioactive elements are mainly in the outer mantle and not the core, which is mainly iron. At the center of the earth there is an iron core that is thought to be solid from measurements of seismic waves that pass through the center.

The continents move about one- half to two inches each

year and their altered positions are the result of Plate Tectonics.

The crust of the continents is lighter than the underlying mantle of the earth and it forms Tectonic Plates of the Lithosphere that "float" above the hotter and lighter mantle. The difference in density of the crust with the mantle is due to the heat of the interior of the Earth. The source of this heat is the decay of radioactive elements inside the earth, which forms the molten shell around the core of the earth with its associated magnetic field. The engine of Plate Tectonics is the creation of new crust from ocean ridges at the junctures of two plates and its spreading motion outward. When Tectonic plates collide, one plate goes under the other and this is called subduction. The driving force of plate subduction is the "wetter", heavier and colder rocks diving into the oceanic ridges (the rocks have a higher percentage of water than those they dive into). Subduction brings new continental material in contact with magma and allows volcanoes to form at these junctures. Therefore, the movement of plates resembles a conveyer belt, with new material coming out at the ridges and then returning into the earth at subduction zones. The collision of plates also triggers earthquakes, Tsunamis, ocean trench formation, and mountain-building. The area surrounding the largest plate, the

Pacific Plate, is known at the "Ring of Fire" for the frequent volcanoes and earthquakes in the northwest U.S., Alaska, Japan and Indonesia. The collision of the Indian and Asian Plates has caused the formation of the Himalayan Mountains. The overall western motion of the plates is thought to be due to tidal forces from the rotation of the moon. Other planets such as Mars have active volcanoes, which are not worn down or changed since plate tectonics does not occur there. In addition, the interior of Mars is no longer molten, its magnetic field is very weak, if it exists at all, and Mars has been stripped of most of its atmosphere by cosmic rays and its weaker gravitational field. The low percentage carbon dioxide atmosphere that it does have is probably due to volcanic activity. The strength of an earthquake is measured by the Richter Seismic Scale. The Richter scale from 1 to more than 10 is logarithmic to encompass the great variation in energy released when the earth's faults move. An earthquake of magnitude 7 is a major earthquake that is keenly felt and ten times more powerful than one of magnitude 6. An earthquake of magnitude 8 causes widespread destruction and loss of life.

Earthquakes cause two kinds of wave in the earth's crust: A pressure wave (P wave) and a transverse wave (S wave). These waves travel throughout the earth and their intensity and timing can be measured with a Seismometer. If the times that these two kinds of waves arrive are recorded, they can accurately pinpoint the location and strength of an earthquake or underground nuclear test. Due to the threat of Tsunamis, which are caused by powerful earthquakes in the ocean, networks of Seismometers line the earthquake areas of the globe. Similarly, networks are located world-wide to detect underground nuclear tests (satellites are in place to detect any above ground or in space).

The current in a conductor is proportional to the voltage and inversely proportional to the resistance in the circuit.
This is Ohm's law, or E = IR, where I is current in amperes, E is the voltage in volts, and R is the resistance in ohms. If there is little resistance in the circuit, say one ohm or less, then for normal house current of 120 volts this means 120 amps of current would pass in the wire. Since power is equal to current squared times resistance (P = I squared times R), the power would be 14,400 watts, resulting in great heat and possibly a fire due to a short circuit. For this reason, household and automobile circuits have fuses that "blow" or become open and do not allow excess current to pass. A 50-amp fuse would have quickly blown under the above conditions, preventing a possible fire.

Matter and Energy in the Universe is neither created nor destroyed.

This is the First Law of Thermodynamics. It states that while matter may be turned into energy and vice-versa, the sum of matter/energy is always the same. Einstein's famous equation E=M times C squared quantifies the equivalence of matter and energy. E is energy, M is mass in kilograms and C is the speed of light in a vacuum. For example, when the nucleus of a uranium atom is split into two halves, two krypton nuclei are formed, and the mass of these nuclei is slightly less than that of the Uranium nucleus. The mass difference between the larger Uranium nucleus and two Krypton nuclei is equivalent to the difference in their respective nuclear binding energies. That mass difference is converted into an enormous quantity of energy.

In any system where work is done or energy exchanged, some of the energy becomes less useable.

This is the Second Law of Thermodynamics. It states that in closed systems, if there is energy or work done there is also an increase in the entropy (a measure of randomness in a system), or the amount of unusable energy. Living things are more ordered but in the process of living more energy is expended than is stored. This law states that there is not 100% efficiency in any mechanical or living process. It also states that the overall entropy (or randomness) of the universe increases with time.

There may be regions in the Universe

where entropy decreases, but that decrease always requires input of energy (for example, solar energy on Earth from the Sun).

The lowest theoretical temperature is 0 degrees Kelvin (-273 degrees Centigrade) and this temperature can never be reached in practice.

This is the Third Law of Thermodynamics. Temperature is simply molecular motion and the scale of temperature can be theoretically extrapolated to a point where such motion would theoretically cease. While physicists have been able to cool liquid helium and other substances within less than a thousandth of a degree of zero degrees Kelvin (called absolute zero), they can never succeed in completely cooling to absolute zero since all atoms intrinsically have some random motion due to their intrinsic energy. However, the Kelvin temperature scale based on this zero point is used by chemists and physicists for measuring and characterizing chemical reactions as well as describing the properties of gases and liquids.

Superconductivity is the complete lack of electrical resistance at very low temperature.

Ordinary conductors such as copper or silver subjected to

temperatures within a few degrees of absolute zero show superconductivity. Electrical resistance in ohms becomes zero and there is not loss of energy in the conduction process. For some special metallic alloys containing copper mixed with other elements, superconductivity occurs at higher temperatures—still very low but close to that of liquid nitrogen. These alloys are being studied to find some optimal alloy that may be commercially viable in electrical transmission over large distances.

Light behaves as a particle in its interaction with matter but light also has characteristics of a wave

Scientists at the turn of the century were puzzled that light's characteristics as a wave could not explain its interactions with metals and other materials. The photoelectric effect is caused when UV light strikes a metal, and a voltage results. It was explained by Einstein, who described how light consists of particles that knock electrons out of the metal to produce the current observed. Physicists later developed Quantum Electrodynamics ("QED") to describe how light interacts with electrons, atoms, and other forms of matter. Though a complex theory, QED was elegantly described by the physicist Richard Feynman in his popular book *Q.E.D: The Strange Story of Light and Matter.* . Q.E.D. theory uses the particle properties of light to explain all of its interactions with atoms and matter.

The primary colors are Red, Orange, Yellow, Green, Blue, Indigo, and Violet ("ROY G. BIV") in order of increasing frequency and energy of the light

The colors of this spectrum may be seen with light refracted by a prism or in a rainbow. Scientists use instruments called spectroscopes to analyze light in the laboratory or from objects in a telescope. The spectra obtained provide information on the composition of the material ionized or of the astronomical objects. For example, ionization (due to high heat or electric current) will cause the element barium to appear green, and strontium red. In a flame, neon is red and mercury blue. Sodium is orange in lamps. Stars and other astronomical objects show spectra characteristic of the elements they contain; gases intervening will block the "lines" specific to the elements absorbing the light.

Astronomers inspect the lines of hydrogen in galaxies to determine their shifting due to velocity, much like sound is shifted in wavelength by the Doppler Effect.

When a light wave encounters a small particle and is scattered, its scattering is in proportion to the fourth power of its frequency.

Conversely, the amount of light passed through a suspension of small particles is inversely proportional to the fourth power of its frequency. For this reason, the sky appears blue; blue light is twice the frequency of red light and therefore is scattered sixteen times as efficiently from the molecules in the sky; the red light mainly passes through. At sunset, light must pass through a great distance from the setting sun close to the horizon. The red light is the only light effectively passing through the length of air and moisture between the sun and the viewer since the shorter wavelengths of light such as blue and green are scattered. Therefore, red is the main color observed at sunset.

Maxwell's First Law of Electromagnetism is known also as Gauss's Law.

Maxwell's four mathematical laws together describe the behavior of electric and magnetic fields and electromagnetic waves (such as light or radio waves). The first law covers the behavior of charges and the electric field. It states that like charges repel and that unlike charges attract, and states that the forces between them are proportional to the inverse square of their separation. This law also states that a charge on an insulated conductor moves to its outer surface.

Maxwell's Second Law of Electromagnetism is also known as Gauss's Law for magnetism.

The mathematical formula in this law describes the magnetic field. It states that it is impossible to create an isolated magnetic monopole. Such a monopole would be a particle that is only a "North" or "South" pole of magnetism and it is deemed impossible by this law.

Maxwell's Third Law of Electromagnetism describes the relation between moving electric charges and their magnetic fields.

This law was originally known as Ampere's Law but was extended by Clerk Maxwell. It states that a current in a wire sets up a magnetic field near the wire. An electric motor can be described by this law since current through its coils create magnetic fields that can drive a rotor made of magnets through the repulsion of like magnetic fields and the attraction of unlike fields. This law also allows one to calculate the speed of light from purely electromagnetic measurements.

Maxwell's Fourth Law of Electromagnetism describes the electrical effects of a changing magnetic field.

The fourth law is also known as Faraday's Law of Induction. It describes how a bar magnet, thrust through a closed loop of wire, will set up a current in the loop. This law therefore covers electrical generators, which use the motion of a rotor turned by rushing water, for example, to cause an alternating current to result in wires in a rotor that is surrounded by a magnetic field. Therefore, this is the basic principle which governs the operation of an electrical generator in a dam.

Planck's Constant quantifies the limits on the sizes of very small subatomic objects or the determination of their velocities and positions.

Planck's constant (h) is a very small number indeed, providing the energy E in a quantum of light or subatomic particle from its frequency v (the Greek letter nu) in the equation E = hv. Planck's constant is 6.8 times 10 to the negative 27th power. It is a unit of action, or energy multiplied by distance. Planck was the first physicist to postulate that there exists a smallest amount of energy, called a quantum. Light and other particles are found in these smallest of packets, called quanta. The product of the velocity or momentum of a particle is always greater than h. This is called the Heisenberg Uncertainty Principle and it comes into play only in the realm of atomic dimensions. For the dimensions we experience in our physical world Planck's constant does not significantly enter calculations since object velocities are too low compared with light or the mass is much larger than that of atomic particles. For example, the Heisenberg Uncertainly Principle says that for the hydrogen atom, we cannot say what position an electron has in its orbit around the nucleus (a proton). The uncertainty in its position is anywhere along the entire radius of its possible orbit around the proton.

Newton's First Law states that "an object at rest tends to stay at rest and an object in motion tends to stay in motion with the same speed and in the same direction unless acted upon by an unbalanced force".

This law says that objects that are at rest tend to stay that way unless acted upon by a force. If they are in motion they tend to move at that speed and direction, unless an outside unbalanced force moves them out of their course. This law describes what we now call inertia. It explains why that drink you are sipping while you are driving your car may spill if you change course (the liquid keeps going in the direction it was moving). It also explains why you may pitch forward when you suddenly put on the brakes: You continue in the direction you were going though the car slows.

Newton's Second Law states that "the acceleration of an object is proportional to the net force and inversely proportional to the mass of the object".

This law provides the equation that the acceleration equals Force divided by the mass. Rearranging this equation gives the more commonly stated form of the law: Force = mass x acceleration. Keep in mind that the force is always the net force, or the vector sum of all the forces acting on the object. Vectors sum means that the direction of the force(s) is always important and must be accounted by the mathematics which analyzes the force depending on its orientation to the object.

Newton's Third Law states that "for every action there is an equal and opposite reaction".

This law describes what occurs when two objects "A" and "B" interact. Object A exerts a force upon object B, but object B exerts an equal and opposite force upon A. A bird flying in the sky flaps its wings against the air; the air exerts an opposite force that lifts the bird off of the ground.

The rate of acceleration due to gravity is 32 feet per second per second (9.8 meters per second per second)

An object under the influence of gravity will reach 32 feet per second at the end of the first second, so it will have gone 16 feet during this time (due to its average velocity of 16 feet per second during that time). However, its fall is slowed by the wind resistance of the air, which according to Newton's Third law is exerting an equal and opposite force that eventually stops the acceleration so that terminal velocity is reached (next topic).

Terminal velocity of the human body in free fall in the atmosphere is 135 miles per hour.

For a person in free fall, the terminal velocity in air is 60 meters per second, or 135 miles per hour. The acceleration of gravity on an object in the earth's gravitational field is 9.8 meters per second per second. Wind resistance of an object counterbalances the force of this acceleration, and that resistance is proportional to the square of the area of the object. So, a sky diver can increase his air speed after reaching terminal velocity by reducing his overall size; the overall velocity may be greatly reduced by deploying the large surface of the parachute. Conversely, a small object such as a bullet has a compact surface area and will reach a terminal velocity greater than 135 miles an hour. For this reason, bullets shot straight into the air may be lethal when they fall back to ground.

Energy of a moving object increases with the square of its velocity.

The energy in a moving object is one-half its velocity squared times the mass. This formula shows that a car at 80 mph has four times the energy of one moving at 40 mph. Similarly, the energy in wind pressing against a surface is proportion to the square of the wind speed (as well as the area of that surface).

Escape velocity from the earth's gravitational field is 11.2 kilometers per second.

The very rapid velocity of 6.7 miles per second is required for an object of any mass to escape the earth's gravitational field. This would be the velocity needed to go to the moon or another planet. To reach low earth orbit a lesser velocity is required, but it is still over 5 miles per second. At the orbital velocity, the kinetic energy of the satellite is equal to the potential energy that it has due to gravity. Therefore, the formula for escape velocity is derived from the equality: ½ mv squared = GMm/r, where G is the gravitational constant, M is the mass of the earth and m the mass of the object and v is the velocity at radius r. Notice that the object's mass cancels out since it is on both sides of the equation. Solving for v provides the escape velocity, v = the square root of 2GM/r.

Red hot metal is at a temperature of 377 to 527 degrees Centigrade.

Any object emits a spectrum of radiation (light) and the maximum wavelength emitted is an exponential function of the fourth power of temperature (Plank's Law). Objects at room temperature emit light in the infrared; this is why an infrared detector will allow objects to be "seen" in the dark. As the temperature is raised, the light frequency shifts higher. Eventually it becomes visible, for example when a steel rod is heated to high temperature between 377 and 527 degrees Centigrade, it glows red (the longest visible wavelength and the least energetic visible light). When heated even higher it becomes "white hot", showing that the light is a mixture of colors from red to yellow to blue, the combination of colors that yields white light. The sun has a surface temperature so hot that much light is emitted in the ultraviolet part of the spectrum. Occasionally, magnetic storms on the sun's surface produce so much energy and are so hot that light is emitted in the X-ray band.

Cosmic Rays increase in intensity with altitude.

Cosmic rays are one of the main natural sources of radiation. These are actually not rays but very high energy particles that shower the earth from outer space and originate outside of the solar system. When they strike the earth's atmosphere, they cause the production of other energetic particles called Pi mesons and Muons. Less frequent but even more energetic are free atomic nuclei that collide with the earth. Fortunately, the earth's atmosphere averages about a kilogram of weight above each square centimeter on earth, and this thickness of material shields most of us from most cosmic rays. However, as one increases in altitude, one receives more cosmic rays. For example, air travelers experience more than twice as many cosmic rays as people at sea level. The main threat of cosmic rays is to astronauts and interplanetary space travelers, where there is virtually no shielding from them. These particles will be the main barrier to safe travel between the planets, as it is very difficult to provide a shield from them. For example, a space vessel would have to have over 200 tons of water in a jacket around it just to shield a few astronauts on board.

The speed of light is 300 million meters per second

The wavelength and frequency of light are related to the velocity through the formula: Velocity = (frequency) x (wavelength). Therefore, one may divide the wavelength (in meters) into the velocity above to obtain frequency and vice versa. By these calculations, a radio station broadcasting at 1500 kilocycles has a wavelength of 200 meters. In Einstein's Theory of Relativity, the speed of light is an absolute maximum for any real thing in the Universe (as opposed to something abstract such a geometrical distribution of things or intersection of things). This absolute maximum helped Einstein develop the Theory of Relativity, which is discussed below. Since the speed of light is the greatest velocity obtainable, it is used to measure the immense distances in the universe through the light year, the distance light travels in a year.

The Special Theory of Relativity shows that objects near the speed of light gain mass and that their local time is slowed.

Einstein's Theory of Special Relativity governs the relationships between space and time and between matter and energy. Both astronomical and physics particle observations have corroborated this theory. One of its primary predictions is that as an object accelerates close to the speed of light it will have a much greater mass. Much of this mass is due to the energy required to reach the velocity near that of light, and since mass and energy are equivalent through the formula E=M(C squared) this extra mass becomes very significant at what are called "Relativistic speeds" (those within a few percent of the speed of light). Another aspect of the equivalence of matter and energy is shown by nuclear energy. When a Uranium 235 atom is split in half by a neutron, the resulting Krypton atoms have slightly less mass than the original Uranium isotope. This small mass corresponds to the particles that kept the larger nucleus intact. That small mass corresponds to an enormous amount of energy which is greater than that found in chemical reactions and is harnessed in nuclear reactors. Another prediction is that at

Relativistic speeds, the time experienced by the object is greatly slowed. A practical result of this "Time dilation" is that the half-life of a short-lived atomic particle such as a Muon increases from 2 microseconds when it is created to 30 milliseconds when it is accelerated within a few percent of light speed.

Einstein's General Theory of Relativity governs non-uniform motion and gravity in the Universe.

Einstein developed the General Theory of Relativity to understand the behavior of non-uniform motion. It elucidates the effects of non-uniform motion such as acceleration. It shows that gravitational and inertial mass are equivalent. It predicts that matter "warps" space-time so that this curvature duplicates the acceleration effects due to gravity. This theory also predicted that light is bent by gravity since light and matter are equivalent. A significant bending of light by very massive objects such as distant galaxies were predicted and subsequently observed in the phenomenon of "gravitational lensing" which is observed through powerful telescopes when a more distant galaxy appears "in front" of a closer galaxy. The altered apparent nearer position of the farther galaxy is caused by the refraction of its light by the nearer galaxy. The General Theory also governs the behavior of light and Black Holes. Black holes are massive stars that have collapsed to a point where light cannot escape from them, and their interiors can not be observed (see article on Black Holes below).

The Universe is about 13.7 billion years old

Direct evidence of the time of origin of the universe comes from the rates of cosmic expansion of the galaxies and from the Cosmic Microwave Background (CMB). It has been known for almost 100 years that all galaxies are apparently expanding away from each other and that the apparent velocity with which they recede is proportional to distance away from the observer. The expansion was first discovered by the astrophysicist Edwin Hubble. Hubble's constant gives distance to a galaxy from its spectral red shift.

Extrapolation back from the expansion implies that the Universe began at one point in time. The light from the most distant galactic objects—Quasars—left those objects more than 12 billion years ago, showing that the universe is at least this old. More precise determination of the age of the Universe comes from calculations from the Cosmic Microwave Background (CMB). Radiation permeates the earth from all directions in uniform strength at a microwave wavelength of about 4 centimeters; this is the CMB. It is the radiation that was released a short period of time after what is called the "Big Bang", the initial explosion that occurred at the beginning of the formation of our universe. The initial temperature was in the thousands of billions of degrees, but that radiation could not be released since it was scattered by the high velocity of the subatomic particles in the universe at those extreme temperatures. After several hundred thousand years, the universe cooled to 3000 degrees K., this radiation was no longer scattered or blocked, and it was released to permeate the universe. It is still everywhere in all directions as direct evidence of the Big Bang. Its initial wavelength was around a micrometer (in the infrared part of the spectrum) but the expansion of the universe to thousands of times in size since then has dramatically shifted its wavelength to about a centimeter, corresponding to objects at only 3 degrees Kelvin. Calculation using the precise fold difference at the time of release and the expansion rate as provided by the Hubble constant yields the present value of 13.7 billion years for the age of the universe.

The expansion of the universe appears to be accelerating.

Astronomers have measured the distances and velocities of other galaxies by determining the brightness of Type IA Supernova, which all reach a certain range of intrinsic brightness during their explosions. This intrinsic brightness enables their distance to be measured from their apparent magnitude measured here on Earth. The data implies that not only is the universe expanding, but its rate of expansion is also increasing. This acceleration of expansion does not appear to have been proceeding since the Big Bang but has occurred more recently in the history of the universe. Physicists have postulated Dark Energy as the cause of this acceleration, but the nature of this dark energy is not understood; nor is it a proven part of astrophysics.

The Solar system is about 4 billion years old.

The age of the solar system has been determined by radioisotope dating of Uranium compared with its decay product Lead in meteorites that are remnants of the formation of the solar system. This meteorite material dates to 4 billion years ago. The oldest fossils of living organisms can be dated by the rocks that contain them and are more than 3.5 billion years old. Other radioactive elements can be used to date the more recent geological past. Potassium-Argon is used for volcanic or other highly heated rocks to determine their age of formation. One out of about 10,000 potassium atoms is naturally potassium-40 (K-40), a radioactive element that becomes Argon-40 when its decay produces a beta ray that strikes a proton, converting it into a neutron. About 11% of decaying K-40 produces Argon-40.

Mass spectrometers can accurately determine the amount of K-40 and Ar-40. From the ratio of these two isotopes and the half-life of K-40, the time when the rock formed can be calculated. When a volcanic rock is formed, Argon is released as a gas and the clock is reset to zero for measuring its age by this method. Rocks can be dated up to four billion years and no less than 100,000 years (due to the errors in the accuracy of measuring small amounts of Ar-40). Living materials can be dated by radiocarbon dating once they have died. While living, they exchange the carbon isotope C-14 with the atmosphere's carbon dioxide, where carbon dioxide containing traces of C-14 isotope was formed by cosmic rays striking the upper atmosphere. Once dead, this exchange stops and the ratio of normal C-12 to C-14 can be used along with the C-14 half-life to accurately determine the age of the fossil or other once-living material.

Most Matter in the Universe is invisible to the eye and is not Baryons or other conventional particles of physics.

Unlike Dark Energy (which is postulated), Dark Matter is a documented scientific fact. Much of the gravitational behavior of ours and other galaxies cannot be explained by the stars, dust, and other matter observable in it. There must be some other matter that can account for the greater gravitational field than that due to conventional matter. Black holes, supermassive particles, and other non-conventional things have been postulated to account for Dark Matter, but the mystery of its identity has not yet been solved.

The Solar Cycle is a 22-year variation in the magnetic, sunspot and storm activity of the sun.

Sunspots are areas of intense magnetic activity on the sun's surface, and the number and placement of the spots varies over a 22-year cycle called the Solar Cycle. Sunspots appear darker than the surrounding solar surface though they are thousands of degrees in temperature. They usually appear in pairs and their position either in the northern or southern solar hemisphere is characteristic of the solar cycle number. The year 2009 began Solar Cycle 24. The magnetic orientation, either north or south is also characteristic of the stage of the solar cycle. The sunspots have intense magnetic fields that emit more particles than the rest of the sun. These particles, called the solar wind, are mainly protons. In addition, sunspots can produce intense geomagnetic storms that affect the propagation of radio waves in the earth's ionosphere as well as the functioning of the electric power grid. The Solar Cycle has a minimum followed by a maximum 11 years later. There are two cycles of minima and maxima in each Solar Cycle, and these are distinguished by the placement and magnetic orientation of sunspots.

Mars is most like the earth in properties that might support life.

Mars has an atmosphere that is primarily carbon dioxide with very little oxygen and a pressure only 1% of that on earth. Mars does have evidence of water. Mars is further from the sun and thus colder (temperature range of -100 deg. F to 20 deg. F.) and it has seasons due to its rotation around the sun. Mars no longer has an active magnetic field, so therefore has less protection from intense cosmic radiation, as well as ultraviolet rays from the sun.

Planets including the Earth have been bombarded by comets and asteroids in the past.

On earth there is abundant evidence of past collisions and cataclysms caused by comet or asteroid impact. The object that collided with the Earth 65 million years ago caused an impact crater that was many miles in width that is now detectable under the earth's surface in the Yucatan. This impact initially caused fires world-wide, then enveloped the earth in smoke and debris, lowering terrestrial temperature, halting much photosynthesis, and resulting in the extinction of the dinosaurs (with the exception of birds, which are the only living descendents of dinosaurs). From July 16 to July 24 in 1994, Comet Shoemaker-Levy collided with Jupiter resulting in a massive storm on that planet, far beyond that which was initially predicted. It astonished the world as it was the first major collision of two bodies in the solar system that has been witnessed. While comets periodically come by the earth, the greatest threat to us is in the form of near-earth asteroids. These asteroids stray from their usual orbit between Jupiter and Mars, and some may come perilously close to the earth.

Supernovas are the massive explosion of stars larger than the sun.

If a star is several times the size of the sun, it burns its hydrogen fuel much faster, and produces more helium and iron. When the mass of these heavier elements increases to the point where fusion can no longer counteract the force of gravity, the whole star collapses, releasing enormous amounts of energy. This is called a Supernova. The collapse results in the formation of an extremely massive and compact core, called a neutron star, since it is comprised of neutrons with a mass less than that which would cause it to collapse further into an object called a Black Hole. This process also results in the release of a huge flux of neutrinos, a by-product of matter being transformed from protons into neutrons. Supernovae shine very brightly, considering their great distance from us. One called 1987A was detected in the twentieth century in our companion galaxy, the Megallanic Cloud. The Crab Nebula is the remnant of a supernova so bright that is was seen both night and day in 1054 A.D.

Black Holes result from the collapse of matter into an object so massive that no light can escape its gravitational field.

Stars or accretions of stars and matter that are above a certain limit in density and mass cause the complete gravitational collapse into an object of unknown size and puzzling properties. At a certain radius, called the Schwartzchild radius, the gravitational field is so strong that it prevents light from escaping and also pulls surrounding matter, gas and particles into it. As the matter accelerates in its fall into the Black Hole, it emits light. This radiation is one of the few ways that Astronomers have of detecting Black Holes, since they emit no light of their own. Enormous Black Holes are thought to reside in the centers of Galaxies, including our own. In some galaxies, black holes are thought to be responsible for enormous jets of particles and energy emitted from their centers in opposite directions.

Pulsars are periodic emitters of radio waves from Neutron Stars.

Astronomers detected sources of radio waves in the sky that vary in their pulsation time from a few milliseconds to many seconds. They determined that these signals coincided with the sites of former Supernovae and are created by the remnant Neutron stars that resulted. The Neutron star emits radio waves in a specific direction due to its intense magnetic field. The spinning of the Neutron star causes us to detect a pulsation as the beam of radio waves rotates to and away from us.

Some Distant Galaxies have Supermassive black holes at their center and emit enormous amounts of radio energy.

These galaxies are located billions of light years from the earth. They were originally called "Quasi-stellar sources" because they appear no larger or brighter than faint stars seen through telescopes. This was abbreviated to "Quasar". A Quasar is formed when very massive black holes form in galaxies; these black holes pull surrounding matter into them so that the matter accelerates and emits radio and other light waves. These are some of the brightest objects in the universe since they can be observed at distances over 12 billion light years. These distances also imply that these objects formed relatively early after the Big Bang.Gamma Ray Bursts are the most violent events that have been detected in the Universe. Gamma Ray Bursts are relatively short—the intense signals lasting a few days, and the physical object associated with it is very hard to detect. But the amount of energy released in these bursts is staggering, exceeding that emitted by supernovae. Gamma Ray bursts have occurred in locations all over the sky and appear to be from distant as well as closer galaxies. They appear to be caused in some cases by a star spinning into a black hole or two black holes spinning into each other. Since their telescope images are very hard to detect in the brief time of their emission, further study of their causes is needed. Two Black Holes combining with each other have recently been detected by the gravitational waves that they emit.

References and Further Reading

Chapter One

1. The Merck Manual of Medical Information, Second Home Edition, Mark H. Beers, ed., Pocket Books, New York, 2003.

2. Biochemistry, Third Edition, Lubert Stryer, W. H. Freeman and Company, New York, 1988.

3. Contemporary Nutrition, Seventh Edition, Gordon Wardlaw and Anna Smith McGraw Hill, New York, 2009.

4. Brock Biology of Microorganisms, Eleventh Edition, Michael T. Madigan and John M. Martinko, Pearson Prentice-Hall, Upper Saddle River, N.J. 2006.
5. Wu, F. *et al* (2020) A new coronavirus associated with human respiratory disease in China. Nature 579: 263-269. http://doi.org/10.1038/s41586-020-2008-3

Chapter Two

1. Foundations of Physics, Eighth edition, David Halliday, Wiley, New York, 2007.

2. Q.E.D. The Strange Theory of Light and Matter, Richard P. Feynman, Princeton University Press, Princeton, N.J., 1988.

Chapter Three

1. Astronomy, Sixth Edition, Eric Chaisson and Steve McMillan, Benjamin-Cummings, New York, 2009.

2. The First Three Minutes. A Modern View of the Origin of the Universe. Steven Weinberg, Bantam Books, New York, 1980.

3. Reading the Rocks: An Autobiography of the Earth, Marcia Bjornerud, Basic Books, New York, 2008.

About the Author

Martin Winkler is a biochemist and consultant in biotechnology. As Principal Scientist at Abbott Laboratories, he helped develop immunodiagnostics for blood borne pathogens such as *Trypanosoma cruzi*, and researched methods for rapid bacterial identification by mass spectrometry. He is the author or co-author of 22 peer reviewed papers and holds four patents.

Made in the USA
Monee, IL
05 November 2026

67962920R00059